Making sure your office complies with the Health
(Display Screen Equipment) Regulations 1992 (as ar

The law on VDUs
An easy guide

HSE BOOKS

© Crown copyright 2003

First published 1994
Second edition 2003

ISBN 0 7176 2602 4

Contents

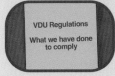
contents

Notes on the second edition

This booklet has been revised to bring it up to date, and we have made some changes in its structure, but the advice it contains is fundamentally the same as that in the first edition.

This guidance is issued by the Health and Safety Executive. Following the guidance is not compulsory and you are free to take other action. But if you do follow the guidance you will normally be doing enough to comply with the law. Health and safety inspectors seek to secure compliance with the law and may refer to this guidance as illustrating good practice.

Introduction

This booklet is for employers who need to comply with the Health and Safety (Display Screen Equipment) Regulations 1992. You may know them as the 'VDU Regulations' or 'Display Screen Regulations.' It is a practical guide, with easy to follow steps, on what to do if you have ordinary office VDUs (visual display units, such as computer screens).

If you have equipment other than office type VDUs which you think may be covered, or if you want more information, you should refer to HSE's detailed guidance booklet L26 which contains the full Regulations (see 'Further publications and sources of advice').

The health problems associated with VDU work are:

■ upper limb disorders (including pains in the neck, arms, elbows, wrists, hands, fingers). Often known as repetitive strain injury or 'RSI';

■ back ache;

■ fatigue and stress;

■ temporary eye strain (but not eye damage) and headaches.

The causes may not always be obvious and can be due to a combination of factors. But enough is known about the importance of some measures - for example, the need to sit properly and take breaks - to allow the risks to be tackled effectively.

The Regulations came into force on 1 January 1993 (some minor changes were made in 2002). They put into UK law a European Community Directive which seeks to protect the health of your workers by reducing risks from VDU work. Briefly, the Regulations require employers to:

■ analyse workstations to assess and reduce risks;

■ ensure that workstations meet specified minimum requirements;

■ plan work activities so that they include breaks or changes of activity;

■ provide eye and eyesight tests on request, and special spectacles if needed;

■ provide information and training.

If your workplace doesn't already comply, you should take action. Following the advice, in the seven steps in this booklet, will help you comply. Each step tells you what you have to do and offers advice about the easiest ways to do it. The checklist on page 21 can be used to assess workstations (with their equipment, furniture and surroundings). It gives solutions to some common problems.

Some of the advice in this guide takes the form of suggestions on how to comply. Where these are steps not actually required by law, this is made clear, eg:

■ 'you **may want** to set a timetable';

■ '**consider** using videos';

■ 'these guidelines **may be helpful**';

■ 'the following points **may help**';

■ 'checklists **are one way** to do assessments'.

Decide who is to be responsible for the following steps.

Step 1:   Decide who is covered by the Regulations

Step 2:   Train users and assessors

Step 3:   Assess workstations and reduce the risks

Step 4:   Make sure workstations and equipment comply with minimum requirements

Step 5:   Plan changes of activity or breaks for users

Step 6:   Provide eye tests and any necessary spectacles for VDU work

Step 7:   Tell users what you have done

You may want to set a timetable for action.

Employees and their safety representatives should be fully involved in implementing the Regulations. They can be particularly useful in providing early warning of health problems, and helping to assess risks, as they have detailed knowledge of day-to-day conditions.

Don't assume you must use consultants to help you comply. You will probably need outside help for eye testing and any investigations of serious aches and pains. But simple problems don't require specialist expertise. Use the in-house knowledge that already exists.

Make sure whoever you choose (it may be more than one person) agrees their responsibilities and knows what is expected of them. Get them to read this booklet. Check later that action has been taken.

Introduction

Step 1

DECIDE WHO IS COVERED BY THE REGULATIONS

The Regulations apply where there are people who 'habitually use display screen equipment as a significant part of their normal work'. So, not everyone who uses a VDU is covered by the Regulations - only those most likely to be at risk. You need to decide who these people are. Remember to include homeworkers and agency 'temps', if you have any (there is more advice on these special groups in our other guidance booklet L26).

People using a VDU more or less continuously on most days will be covered by the Regulations. So, usually, are others who:

- normally use a VDU for **continuous or near-continuous spells of an hour or more at a time**; and

- use it in this way **more or less** daily; and

- have to **transfer information quickly** to or from the display screen equipment;

and also need to apply high levels of **attention and concentration**; or are **highly dependent** on VDUs to do the job or have **little choice** about using them; or need **special training or skills** to use the equipment.

Such people are called users in this guidance. Users may include, for example, people who do word processing or data input, secretaries, telesales personnel, journalists, librarians, graphic designers, and many others.

The Regulations distinguish between employees (**users**) and self-employed workers (**operators**). This distinction is not used here - where certain obligations do not apply to the self-employed, this is made clear in the text.

Who are your VDU users?

Have you identified them all?

Step 1

Step 2

TRAIN USERS AND ASSESSORS

Arrange training for:

■ users on the risks, and safe behaviour and practices. For example, adjustable chairs only reduce risk if users know how to adjust them and sit properly. If big changes are made to workstations, users may need retraining;

■ workstation assessors. Step 3 requires you to assess workstations. Assessors will need to recognise risky workstation layouts, environments and practices. You can train your own staff to do this job.

Good user training should normally cover:

■ the risks from DSE work (see Introduction);

■ the importance of good posture and changing position;

■ how to adjust furniture to help avoid risks;

■ organising the workplace to avoid awkward or frequently repeated stretching movements;

■ avoiding reflections and glare on or around the screen;

■ adjusting and cleaning the screen and mouse;

■ organising work for activity changes or breaks if necessary;

■ who to contact for help and to report problems or symptoms;

■ contributing to the risk assessment, eg completing checklists.

You do not have to give health and safety training to self-employed people using your workstations. That is their responsibility.

Consider using:

■ videos;

■ computer-based training;

■ the HSE leaflet *Working with VDUs* (see 'Further publications and sources of advice');

■ wall charts;

■ seminars.

Make sure users can ask about points that aren't clear. For example, if you show a video, allow time for questions at the end and have someone present who will know the answers.

Good training for workstation assessors will cover the points above, plus:

■ how to review checklists;

■ how to identify obvious and less obvious hazards;

■ deciding when additional information and help is needed, and where to go for it;

■ how to draw conclusions from assessments and identify steps to reduce risks;

■ recording problems;

■ how to tell those who need to take action on findings, and give feedback to users.

Methods of training assessors include:

■ professionally arranged seminars;

■ getting familiar with HSE guidance *Display screen equipment work: Guidance on Regulations* (see 'Further publications and sources of advice');

■ computer-based training.

Whatever training methods you use, you should check afterwards that assessors have understood the information and have reached an adequate level of competence.

Step 3

ASSESS WORKSTATIONS AND REDUCE THE RISKS

Checklists (like the one on page 21) are one easy way to do workstation assessments, and enable users to take an active part. After training, users can fill in the checklist themselves. They know what the problems are, and whether or not they are comfortable. For example, a workstation assessor could assess a workstation in the morning and find no glare on the screen; only the user would know that glare is bad in the afternoon.

Remember, the user filling in the checklist is only the first stage, not the whole assessment. **A properly trained assessor should go over the completed checklists, clarify any doubtful points, and tackle problems that the user can't solve.**

Make arrangements to review assessments when there is a significant change to the workstation, for example when it is relocated or a different screen is installed.

Users can answer the questions in the first column of the checklist, making adjustments as they go, to reduce any problems they find. Where they answer 'Yes', no further action is necessary.

Standard workstation items (for example display screens, keyboards, chairs) do not need individually assessing but users will need to check that their items function properly. For example, if all chairs are the same make and model you will know whether they have an adjustment mechanism, but users will need to check that their own chair mechanism works.

Assessors may find the following guidelines helpful:

■   deal with the biggest problems first;

■   take seriously and investigate reports of aches and pains from users;

■ look for the less obvious causes of risk. For example, poor (ie risky) posture may be due to bad seating, **or** sitting awkwardly to avoid glare on the screen, **or** leaning forward to key because arm rests prevent the chair being close to the workstation, **or** a poorly positioned mouse;

■ remember to assess all the risks - look at things like task demands and rest breaks, as well as the physical aspects of the workstation;

■ take account of any special needs of individuals, such as users with a disability;

■ consider different ways of tackling risks, eg if keyboard and screen are fixed (as they are with many portables), risks could be reduced by increasing job variety, for example, by including more time working away from the computer;

■ check later to see that action identified by the assessment has in fact been taken, and that it is not causing other problems (see page 10);

■ sign-off the checklist on its first page to show that everything has been done.

Beware of misinformation and exaggerated claims from suppliers of products that are supposed to reduce risks. They may oversimplify issues or talk up health 'risks' to persuade people to buy accessories. Not everything advertised as 'ergonomic' has been properly tested and found to be helpful. Beware in particular the following, which are **unnecessary**:

■ tinted VDU spectacles;

■ over-elaborate adjustable tables;

■ radiation-reducing devices (filters, lead aprons).

You do not have to use checklists to do assessments, but many people find them useful.

Step 3

Having taken action to reduce the risks, check with the user that no new problems have arisen, for example:

■ to have forearms in the correct keying position, a short user raises the chair height, but feet can't now be placed flat on the floor. A footrest is needed;

■ workstation layout is reorganised to give more space but one user is now sitting next to a source of noise or glare.

Completed assessments will need to be reviewed when:

■ major changes are made to the equipment, furniture, work environment or software;

■ new users start work, or change workstations;

■ workstations are relocated;

■ the nature of work tasks changes considerably.

Focus on the aspects that have changed. For example:

■ the environmental factors are important if the workstation location changes;

■ different users have different needs - replacing a tall person with a short one may mean a footrest is required;

■ users working from a number of source documents need more desk space than users who just use the computer and telephone.

Use your assessment to decide what needs to be done. Check that action has reduced the risk

Step 3

Step 4

MAKE SURE WORKSTATIONS AND EQUIPMENT COMPLY WITH MINIMUM REQUIREMENTS

The Schedule to the Regulations sets out some minimum requirements for ergonomic features that workstations should have (whether or not they are used by a user). You can use the checklist on page 21 to help you comply. Manufacturers and suppliers can assist, but remember that you as the employer have the duty to ensure items comply.

The Schedule covers broad design factors for furniture; the VDU hardware, software and accessories; and the workstation environment. It applies to equipment bought second-hand and new. It does not include detailed measurements and specifications. There is no requirement for equipment to comply with British or international standards. (However, choosing things that do comply with relevant standards - such as BS EN ISO 9241 - can help make sure they will satisfy or go beyond the requirements in the Schedule.)

The questions and 'Things to consider' in the checklist cover the requirements of the Schedule. If you can answer 'Yes' in the second column against all the questions, having taken account of the 'Things to consider', you are complying. You will not be able to address some of the questions and 'Things to consider', eg on reflections on the screen, or the user's comfort, until the workstation has been installed. These will be covered in the risk assessment that you have to do once the workstation is installed. Follow Step 3 to complete the assessment and reduce any risks you find.

You don't need to take measurements of furniture, temperature, noise levels, radiation, etc to comply with the schedule. Adjustable tables are not a legal requirement.

Note that because of the way the Schedule is worded, you don't have to comply with all the detailed requirements in the checklist if:

■ the workstation doesn't have a particular item.

For example, some kinds of display screen work do not need a document holder. Complying with the Schedule doesn't mean you always have to provide a document holder;

■ it would not be helpful to comply with one (or more) of the requirements as this would not improve health and safety in a particular case.

For example, some users may have certain back complaints that benefit from having a chair without a backrest. In this case providing a chair with a backrest could do more harm than good;

■ the nature of the task makes it inappropriate.

For example to preview page layout on screen, it may not be necessary for characters to be well defined because reading the words is not the priority.

Many workstations have a mouse, trackball or other **pointing device.** Advice on things to look for to ensure users' health is included in the checklist.

Portable computers are subject to the Regulations if they are in prolonged use. You should give their users adequate training. Unless you have made other arrangements, this should include training them to make a risk assessment when setting up a portable away from the office. Risk assessments should consider potential risks from manual handling, if portable computer users have to carry heavy equipment and papers. You should also consider possible risks

from violent theft, as well as the normal risks of all display screen work described in Step 3.

Design features of portables can lead to postural and other problems. To avoid these, tell users to take more frequent breaks, and avoid using the portable in locations such as a car, which are not ideal. Whenever possible, users should be encouraged to use their portable at a docking station. Or, they can use the portable with a plug-in full-sized keyboard and mouse, enabling the height and position of the portable's screen to be adjusted by placing it on raiser blocks.

When selecting a portable, look for low weight (including any accessories) and a large clear screen which can be used comfortably for the tasks to be done. Other ergonomic features may be required for the tasks to be done (for example, equipment for use outdoors needs to be waterproof, with a screen that can be read in bright sun).

Step 5

PLAN CHANGES OF ACTIVITY OR BREAKS FOR USERS

Breaking up long spells of VDU work helps prevent fatigue, eye strain, upper limb problems and back-ache. Where possible, include spells of other work, eg telephone calls, filing, photocopying etc. If such changes of activity are not possible the law requires you to plan for users to take rest breaks.

When organising users' work the following points may help:

- vary the tasks, eg encourage users to walk across to a colleague to get information, rather than e-mailing or using the phone;

- remind users to stretch and change position;

- encourage users to look into the distance from time to time, and to blink often;

- breaks should be taken before users get tired, rather than to recover;

- short frequent breaks are better than longer, infrequent ones;

- individual control over work patterns is the ideal - but make sure users don't:

  - get carried away and work intensely for too long;

  - save breaks to take a few longer ones or go home early; or

  - use breaks for other computer activities, like surfing the Internet.

■ imposed rest breaks may sometimes be the only solution, eg in some data preparation or call centre work;

■ breaks should be taken away from the screen if possible.

Timing and length of breaks is not set down in the law. Needs vary depending on the work done. You are not responsible for providing breaks for the self-employed.

Step 5

Step 6

PROVIDE EYE TESTS AND ANY NECESSARY SPECTACLES FOR VDU WORK

If you employ users, or those about to become users, they can request an eye and eyesight test that you have to pay for. If the test shows they need glasses specifically for their VDU work, you have to pay for a basic pair of frames and lenses.

Users are entitled to further tests at regular intervals after the first test, and in between if they are having visual difficulties which may reasonably be considered to be caused by their VDU work.

You can decide what arrangements to make to provide eye and eyesight tests. For example, some employers let users arrange tests for themselves (and hand the employer the bill); others prefer to send all their staff to be tested by one optician.

When setting up a system for providing users with eye and eyesight tests, these points might help:

■ contact a number of opticians and ask what they charge for tests and basic glasses;

■ ask if they will come to the firm to test users;

■ ask for standard information about each user they test (ie if they need glasses for VDU work, and when they should be re-tested);

■ tell users what arrangements you have made;

■ make sure users understand what you will and won't pay for (eg tinted lenses, or glasses not required for the VDU work are not your responsibility).

VDU work does not cause permanent damage to eyes or eyesight. Eye tests are provided to ensure that users can comfortably see the screen, to work effectively without visual fatigue.

If users' normal glasses for other work are suitable for VDU work you don't need to pay for them. You don't have to pay for fancy frames or lenses. Eye tests are not an entitlement for the self-employed.

Step 6

Learning Resources Centre

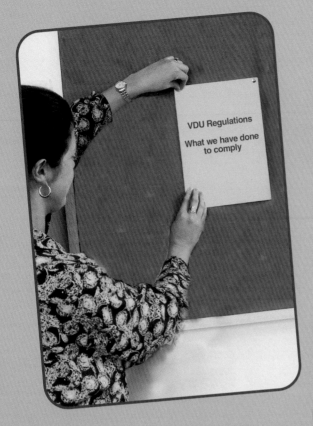

Step 7

TELL USERS WHAT YOU HAVE DONE

Give users information on:

■ health and safety relating to their workstations;

■ risk assessment and steps taken to reduce risks;

■ breaks and changes of activity;

■ eye and eyesight tests.

For self-employed workers you do not need to cover the last two points

You can pass on the required information by:

■ telling staff;

■ putting information in staff instructions on health and safety;

■ circulars;

■ wall charts;

■ computer-based information systems (if staff are trained and can use them).

Make sure you have told your VDU users about the issues relevant to their health and safety

19

The law on VDUs: An easy guide

VDU WORKSTATION CHECKLIST

Workstation location and number (if applicable): ....

User: ....

Checklist completed by: ....

Assessment checked by: ....

Date of assessment: ....

Any further action needed? YES/NO

Follow-up action completed on: ....

This checklist can be used as an aid to risk assessment and to help comply with the Schedule to the Health and Safety (Display Screen Equipment) Regulations.

Work through the checklist, ticking either the 'yes' or 'no' column against each risk factor:
- 'Yes' answers require no further action.
- 'No' answers will require investigation and/or remedial action by the workstation assessor. They should record their decisions in the 'Action to take' column. Assessors should check later that actions have been taken and have resolved the problem.

Remember the checklist only covers the workstation and work environment. You also need to make sure that risks from other aspects of the work are avoided, for example by giving users health and safety training, and providing for breaks or changes of activity. Advice on these is given in the main text of the guidance.

VDU workstation checklist

RISK FACTORS	Tick answer		THINGS TO CONSIDER	ACTION TO TAKE
	YES	NO		

1 Display screens

RISK FACTORS	Tick answer		THINGS TO CONSIDER	ACTION TO TAKE
Are the characters clear and readable?			Make sure the screen is clean and cleaning materials are made available. Check that text and background colours work well together.	
Is the text size comfortable to read?			Software settings may need adjusting to change text size.	
Is the image stable, ie free of flicker and jitter?			Try using different screen colours to reduce flicker, eg darker background and lighter text. If problems still exist, get the set-up checked, eg by the equipment supplier.	
Is the screen's specification suitable for its intended use?			For example, intensive graphic work or work requiring fine attention to small details may require large display screens.	
Are the brightness and/or contrast adjustable?			Separate adjustment controls are not essential, provided the user can read the screen easily at all times.	
Does the screen swivel and tilt?			Swivel and tilt need not be built in; you can add a swivel and tilt mechanism. However, you may need to replace the screen if: • swivel/tilt is absent or unsatisfactory; • work is intensive; and/or • the user has problems getting the screen to a comfortable position.	
Is the screen free from glare and reflections?			Use a mirror placed in front of the screen to check where reflections are coming from. You might need to move the screen or even the desk and/or shield the screen from the source of reflections. Screens that use dark characters on a light background are less prone to glare and reflections.	
Are adjustable window coverings provided and in adequate condition?			Check that blinds work. Blinds with vertical slats can be more suitable than horizontal ones. If these measures do not work, consider anti-glare screen filters as a last resort and seek specialist help.	

RISK FACTORS	Tick answer		THINGS TO CONSIDER	ACTION TO TAKE
	YES	NO		

2 Keyboards

RISK FACTORS			THINGS TO CONSIDER	ACTION TO TAKE
Is the keyboard separate from the screen?			This is a requirement, unless the task makes it impracticable (eg where there is a need to use a portable).	
Does the keyboard tilt?			Tilt need not be built in.	
Is it possible to find a comfortable keying position?			Try pushing the display screen further back to create more room for the keyboard, hands and wrists. Users of thick, raised keyboards may need a wrist rest.	
Does the user have good keyboard technique?			Training can be used to prevent: • hands bent up at wrist; • hitting the keys too hard; • overstretching the fingers.	
Are the characters on the keys easily readable?			Keyboards should be kept clean. If characters still can't be read, the keyboard may need modifying or replacing. Use a keyboard with a matt finish to reduce glare and/or reflection.	

3 Mouse, trackball etc

RISK FACTORS			THINGS TO CONSIDER	ACTION TO TAKE
Is the device suitable for the tasks it is used for?			If the user is having problems, try a different device. The mouse and trackball are general-purpose devices suitable for many tasks, and available in a variety of shapes and sizes. Alternative devices such as touchscreens may be better for some tasks (but can be worse for others).	
Is the device positioned close to the user?			Most devices are best placed as close as possible, eg right beside the keyboard. Training may be needed to: • prevent arm overreaching; • tell users not to leave their hand on the device when it is not being used; • encourage a relaxed arm and straight wrist.	

RISK FACTORS	Tick answer		THINGS TO CONSIDER	ACTION TO TAKE
	YES	NO		
Is there support for the device user's wrist and forearm?			Support can be gained from, for example, the desk surface or arm of a chair. If not, a separate supporting device may help. The user should be able to find a comfortable working position with the device.	
Does the device work smoothly at a speed that suits the user?			See if cleaning is required (eg of mouse ball and rollers). Check the work surface is suitable. A mouse mat may be needed.	
Can the user easily adjust software settings for speed and accuracy of pointer?			Users may need training in how to adjust device settings.	

4 Software

Is the software suitable for the task?			Software should help the user carry out the task, minimise stress and be user-friendly. Check users have had appropriate training in using the software. Software should respond quickly and clearly to user input, with adequate feedback, such as clear help messages.	

5 Furniture

Is the work surface large enough for all the necessary equipment, papers etc?			Create more room by moving printers, reference materials etc elsewhere. If necessary, consider providing new power and telecoms sockets, so equipment can be moved. There should be some scope for flexible rearrangement.	
Can the user comfortably reach all the equipment and papers they need to use?			Rearrange equipment, papers etc to bring frequently used things within easy reach. A document holder may be needed, positioned to minimise uncomfortable head and eye movements.	
Are surfaces free from glare and reflection?			Consider mats or blotters to reduce reflections and glare.	

RISK FACTORS	Tick answer		THINGS TO CONSIDER	ACTION TO TAKE
	YES	NO		
Is the chair suitable? Is the chair stable? Does the chair have a working: ● seat back height and tilt adjustment? ● seat height adjustment? ● swivel mechanism? ● castors or glides?			The chair may need repairing or replacing if the user is uncomfortable, or cannot use the adjustment mechanisms.	
Is the chair adjusted correctly? 			The user should be able to carry out their work sitting comfortably. Consider training the user in how to adopt suitable postures while working. The arms of chairs can stop the user getting close enough to use the equipment comfortably. Move any obstructions from under the desk.	
Is the small of the back supported by the chair's backrest?			The user should have a straight back, supported by the chair, with relaxed shoulders.	
Are forearms horizontal and eyes at roughly the same height as the top of the VDU?			Adjust the chair height to get the user's arms in the right position, then adjust the VDU height, if necessary.	
Are feet flat on the floor, without too much pressure from the seat on the backs of the legs?			If not, a foot rest may be needed.	

RISK FACTORS	Tick answer		THINGS TO CONSIDER	ACTION TO TAKE
	YES	NO		

6 Environment

RISK FACTORS	YES	NO	THINGS TO CONSIDER	ACTION TO TAKE
Is there enough room to change position and vary movement?			Space is needed to move, stretch and fidget. Consider reorganising the office layout and check for obstructions. Cables should be tidy and not a trip or snag hazard.	
Is the lighting suitable, eg not too bright or too dim to work comfortably?			Users should be able to control light levels, eg by adjusting window blinds or light switches. Consider shading or repositioning light sources or providing local lighting, eg desk lamps (but make sure lights don't cause glare by reflecting off walls or other surfaces).	
Does the air feel comfortable?			VDUs and other equipment may dry the air. Circulate fresh air if possible. Plants may help. Consider a humidifier if discomfort is severe.	
Are levels of heat comfortable?			Can heating be better controlled? More ventilation or air-conditioning may be required if there is a lot of electronic equipment in the room. Or, can users be moved away from the heat source?	
Are levels of noise comfortable?			Consider moving sources of noise, eg printers, away from the user. If not, consider soundproofing.	

7 Final questions to users...

- Ask if the checklist has covered all the problems they may have working with their VDU.

- Ask if they have experienced any discomfort or other symptoms which they attribute to working with their VDU.

- Ask if the user has been advised of their entitlement to eye and eyesight testing.

- Ask if the user takes regular breaks working away from VDUs.

Write the details of any problems here:

Further publications and sources of advice

Publications

- *Aching arms (or RSI) in small businesses: Is ill health due to upper limb disorders a problem in your workplace?* Leaflet INDG171(rev1) HSE Books 2003 (single copy free or priced packs of 15 ISBN 0 7176 2600 8)

- *Understanding ergonomics at work: Reduce accidents and ill health and increase productivity by fitting the task to the worker* Leaflet INDG90(rev2) HSE Books 2003 (single copy free or priced packs of 15 ISBN 0 7176 2599 0)

- *Work with display screen equipment. Health and Safety (Display Screen Equipment) Regulations 1992.Guidance on Regulations* L26 HSE Books 2003 ISBN 0 7176 2582 6

- *Working with VDUs* Leaflet INDG36(rev1) HSE Books 1998 (single copy free or priced packs of 10 ISBN 0 7176 1504 9)

- *Work-related stress: A short guide* Leaflet INDG281(rev1) HSE Books 2001 (single copy free or priced packs of 10 ISBN 0 7176 2112 X)

Further copies of the checklist in this guide are available: *VDU workstation checklist* Leaflet HSE Books (priced packs of 5 ISBN 0 7176 2617 2)

Advice from Health and Safety Authorities:

- For businesses in office or retail premises, contact the Environmental Health Department at the local Council.

- For other premises, contact the nearest Health and Safety Executive Regional Office, as listed in Yellow Pages.

Learning Resources Centre

Printed and published by the Health and Safety Executive   C75   02/03